LONGEST TALLEST HEAVIEST

Contents

Long lizards	2
Standing tall	3
Small and swift	4
Big birds	5
Sound asleep	6
Fast finish	7
Living legends	8
Heavy loads	9
Walking tall	10
Hold your breath	11
Lightweights	12
Insect giants	13
Flying high	14
Laying an egg	15
Glossary	16
Index	*Inside back cover*

Calvin Irons

Long lizards

DINOSAURS ARE the largest *animals* ever to walk the earth. The word **dinosaur** means **terrible lizard**.

This complete skeleton of the **diplodocus** (die-PLOD-o-kus) is 90 feet long. Other dinosaurs were probably longer, but so far, only a few of their bones have been found.

Diplodocus was about the same *length* as 9 small cars.

Longest dinosaurs
(Estimates from bones found)

Seismosaurus	120 feet
Supersaurus	100 feet
Antarctosaurus	95 feet
Diplodocus	90 feet
Brachiosaurus	80 feet

1. Find the length of one of the cars. How did you figure it out?

2. How many cars would fit along
   - seismosaurus?
   - supersaurus?
   - brachiosaurus?

3. Measure your *armspan*. *Estimate* the length of diplodocus in armspans.

Standing tall

EVERYONE KNOWS that the giraffe is the tallest animal. You might be surprised at how tall some other animals are.

The **second tallest** animal is the African elephant. Males (or bull elephants) are much taller than females.

The grizzly bear is the **third tallest** animal. Other bears are also tall when they stand on their hind legs.

1. Estimate the *height* of each of the animals in the picture.

2. How much taller is the giraffe than each of the other animals? Explain how you estimated each *difference*.

3. How much taller than the man is each of the animals in the picture? Explain how you estimated each difference.

Small and swift

NOT ALL DINOSAURS were large and slow. The smallest dinosaur was the size of a chicken and could run very fast. It was called **compsognathus** (komp-SOG-na-thus). *Fossils* show that its *skeleton* was very similar to that of the first *bird*, **archaeopteryx** (AR-kee-OP-ter-iks).

Lengths of small dinosaurs

Avimimus	60 in.
Compsognathus	24 in.
Heterodontosaurus	48 in.
Hypsilophodon	90 in.
Lesothosaurus	40 in.

1. Give the length of each small dinosaur using
   - feet and inches
   - yards and inches.
2. Calculate the difference in length between compsognathus and each of the other small dinosaurs.
3. Measure your *cubit*. How does it compare to the length of compsognathus?

Big birds

MOST BIG BIRDS have long skinny legs. Some of them can fly, but the biggest birds are too heavy to get off the ground!

Yards 3

The **ostrich** is the tallest, heaviest bird alive. It cannot fly.

The New Zealand **moa** was a giant flightless bird. It became *extinct* about 300 years ago.

2

1

The tallest flying bird is the **sarus crane**.

1. How much shorter than the moa is
   - the ostrich?
   - the crane?
2. Which big bird's height is closest to
   - 3 yards?
   - $2\frac{1}{2}$ yards?
   - 2 yards?
   - $1\frac{1}{2}$ yards?
   - 1 yard?
3. Which two birds have a difference in height of approximately
   - 1 yard?
   - $1\frac{1}{2}$ yards?

0

Heights of some big birds

Flightless birds		Flying birds	
Moa	118 in.	Crane	63 in.
Ostrich	94 in.	Flamingo	59 in.
Emu	71 in.	Condor	55 in.
Rhea	59 in.	Heron	51 in.
Penguin	47 in.	Stork	43 in.

Sound asleep

SOME ANIMALS SLEEP all winter. Other animals are sleepy every day of the year. The clock graphs on this page show the total number of hours per day each animal sleeps.

Koala
The koala sleeps for 22 hours each day! It is awake for only 2 hours.

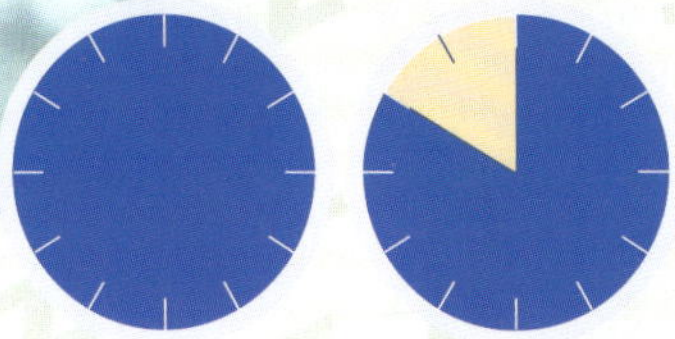

Sloth
The sloth is a really slow mover. It sleeps hanging upside down.

Opossum
The opossum only wakes in the cool of the night.

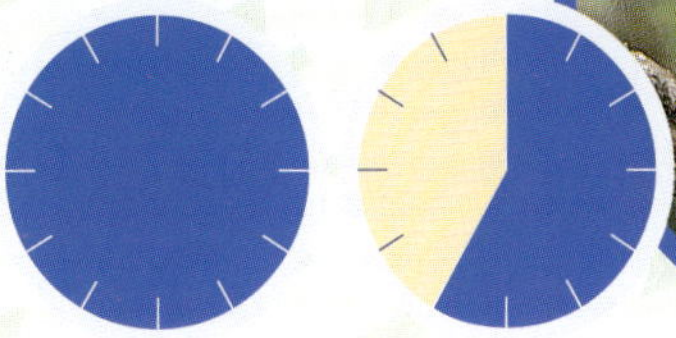

Hamster
The hamster spends more than half the day curled up asleep.

1. Choose one of the sleepy animals. Give the number of hours each day
   - it sleeps
   - it is awake.
2. How many hours do you sleep in one day?
3. For how many hours more than you does each animal sleep?

Fast finish

THE FASTEST ATHLETES can run 100 meters in less than 10 seconds. But many animals would cross the finish line way ahead of them!

Time taken to travel 100 meters

Some fast animals

Cheetah	4 seconds
Antelope	5 seconds
Giraffe	8 seconds

Some slow animals

Snail	2 hours
Sloth	30 minutes
Tortoise	15 minutes

1. Choose one of the **fast** animals in the chart. Estimate how far it would run in **10 seconds**. How did you make your estimate?

2. Choose one of the **slow** animals. Estimate how far it would crawl in **3 hours**. How did you make your estimate?

Living legends

THE LARGEST LAND ANIMAL today is the African elephant. However, African elephants are nowhere near as large as the biggest dinosaurs were.

African elephant

It would take 100 adults to *balance* the weight of one elephant.

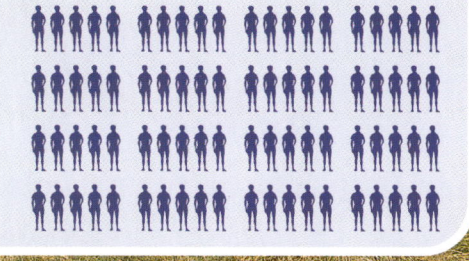

White rhinoceros

The white rhinoceros is much larger than any human, but only half the size of the African elephant.

Ostrich

The heaviest bird is the ostrich. It weighs as much as two adults. The ostrich is so heavy it cannot fly.

1. How many adults would it take to balance the white rhinoceros?
2. How many ostriches would balance:
   - 20 adults?
   - one African elephant?
   - one white rhinoceros?

Heavy loads

THE BRACHIOSAURUS (BRAK-ee-o-SOR-us) was the heaviest dinosaur. It would have taken 12 elephants to balance one brachiosaurus.

The heaviest animals today live in the ocean. The blue whale is the giant of the living world.

Whale Weights *(Measured in Elephants)*

Blue whale
🐘🐘🐘🐘🐘🐘
🐘🐘🐘🐘🐘🐘
🐘🐘🐘🐘🐘🐘

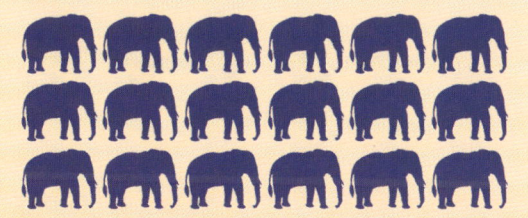

Right whale
🐘🐘🐘🐘🐘🐘

Grey whale
🐘🐘🐘🐘🐘

Humpback whale
🐘🐘🐘🐘🐘

1. Give the difference between the *weight* of the brachiosaurus and each type of whale. (Use elephants in your answers!)

2. How many adults would be needed to balance
   • the right whale?
   • the gray whale?
   • the humpback whale?
   (Use the information on page 8 to find the answer.)

Walking tall

HUMANS, MONKEYS, AND APES belong to a group of *mammals* called *primates*. Like humans, primates have hands with five fingers and can stand on their hind feet.

Gorilla

Gorilla records
The gorilla is the largest primate. The heaviest gorilla ever recorded weighed 680 pounds. The tallest gorilla was 75 inches high.

Orang-utan

The name of this ape means **wild man of the woods.**

Usual size of large primates
Weight and height (or length)*

Gorilla	480 lbs.	70 in.
Orangutan	160 lbs.	65 in.
Chimpanzee	160 lbs.	50 in.
Baboon	110 lbs.	40 in.
Probocis monkey	55 lbs.	30 in.
Hanuman langur	45 lbs.	40 in.
Siamung gibbon	25 lbs.	35 in.

*not including tail

1. Compare the gorilla records with the weight and height chart. What do you notice?

2. Choose two primates from the chart and compare their weights. Then compare their lengths.

3. Find the primates that are heavier than you. How much heavier are they?

4. Find the primates that are taller than you. How much taller are they?

Hold your breath

FISH CAN BREATHE under water, but many aquatic animals need to come up for air. Some of them can hold their breath for much longer than humans can.

Scientists have timed some very long dives. One **elephant seal** stayed under water for 119 minutes. A **sperm whale** held its breath for 112 minutes.

Hippopotamus

A hippopotamus can stay under water for more than 30 minutes.

Underwater times

Minutes

Crocodile	20	Penguin	18
Dolphin	15	Platypus	11
Hippo	30	Seal	119
Otter	4	Whale	112

1. Which animals hold their breath for less than
   - 1 hour?
   - $\frac{1}{2}$ hour?
   - $\frac{1}{4}$ hour?

2. Find the difference between each time in the chart and
   - 1 hour
   - $\frac{1}{2}$ hour
   - $\frac{1}{4}$ hour.

Lightweights

IF ONE OF the two *lightweights* on this page perched on the end of your finger, you might not even know it was there!

Pygmy shrew

A shrew is a tiny mouselike animal. The pygmy shrew is the world's smallest and lightest **mammal**. It comes from the Mediterranean region.

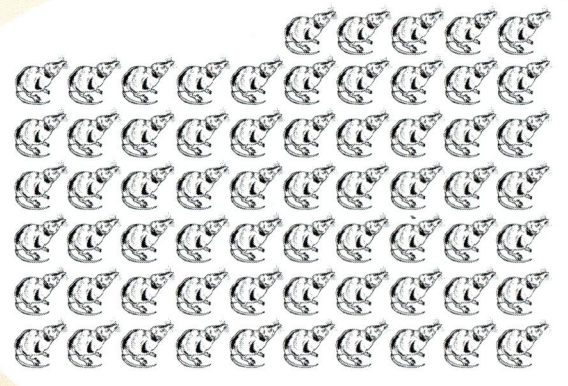

It would take **65** pygmy shrews to balance one small apple.

Bee hummingbird

The world's smallest **bird** is the bee hummingbird. It comes from the West Indies.

You would need **6** bee hummingbirds to balance one strawberry.

1. Suppose a small apple weighs 4 ounces. How many pygmy shrews would you need to balance
   - $\frac{1}{2}$ pound?
   - 1 pound?

2. Suppose a strawberry weighs $\frac{1}{4}$ ounce. How many bee hummingbirds would you need to balance
   - 2 ounces?
   - 4 ounces?
   - 1 pound?

Insect giants

MOST PEOPLE THINK that *insects* are small. But the giants of the insect world are truly colossal.

Stick insect

The **longest** insect on record is a stick insect from Indonesia. It measures 21.8 inches!

Giant hercules moth

The giant hercules moth from Australia has a **wingspan** of up to 11.8 inches.

Goliath beetle

The African goliath beetle is the **heaviest** insect. It weighs up to 3.5 ounces.

1. Choose one of the lengths on this page. How much more or less is it than 1 foot?

2. Measure your *handspan*. Estimate these lengths in handspans:
   - the stick insect
   - the wingspan of the giant hercules moth

3. Would four goliath beetles weigh more or less than 1 pound? Explain your thinking.

Flying high

FLYING *REPTILES* were called **pterosaurs** (TER-o-sors). Their wings were made of skin, and they did not have any feathers.

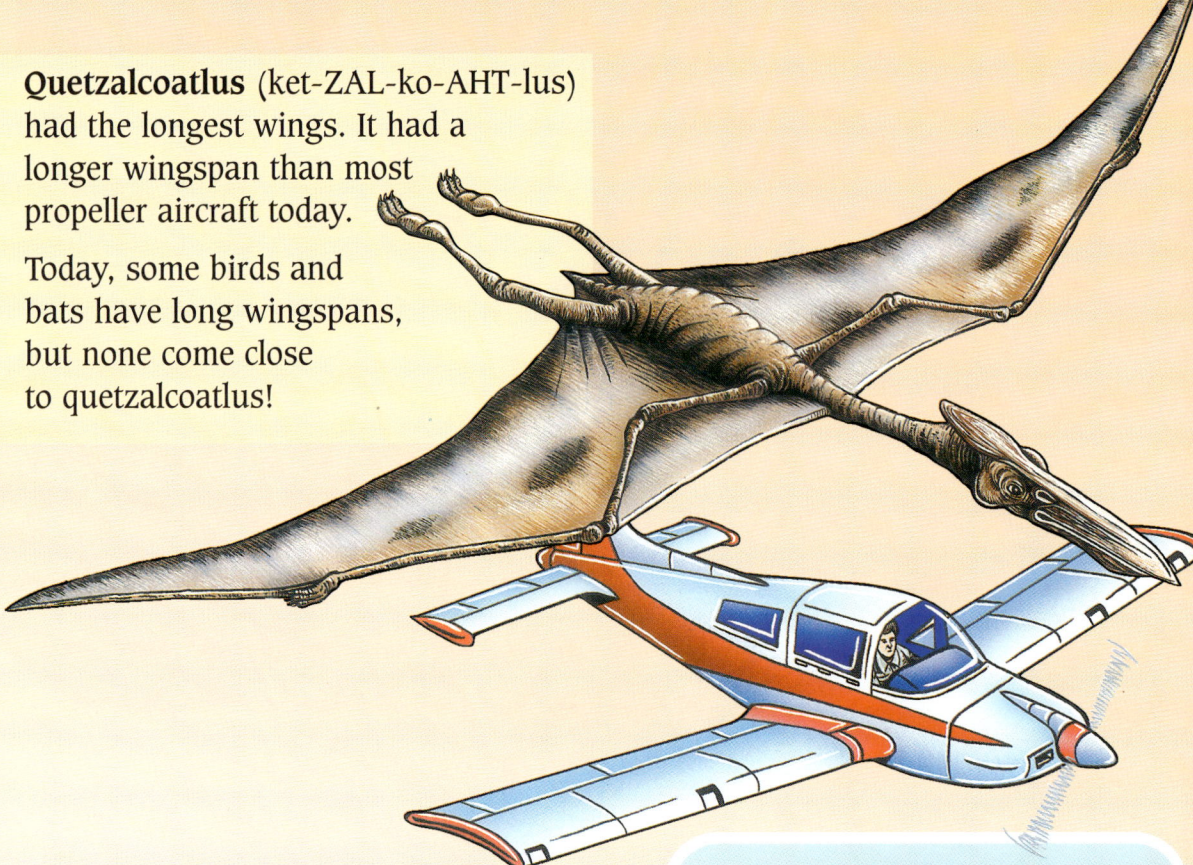

Quetzalcoatlus (ket-ZAL-ko-AHT-lus) had the longest wings. It had a longer wingspan than most propeller aircraft today.

Today, some birds and bats have long wingspans, but none come close to quetzalcoatlus!

Longest wingspans

Quetzalcoatlus	13.1 yards
Marabou stork	4.4 yards
Albatross	4.0 yards
Trumpeter swan	3.7 yards
Kalong *(fruit bat)*	1.9 yards

1. Find the difference between the wingspan of quetzalcoatlus and each of the other wingspans in the chart. Give each answer in yards.

2. Find a pair of wingspans with a difference **less than** 1 yard. Give the difference in inches.

Laying an egg

BIRDS' EGGS come in many different sizes. The ostrich lays the largest egg, and the bee hummingbird lays the smallest egg. A hen's egg weighs less than 2 ounces.

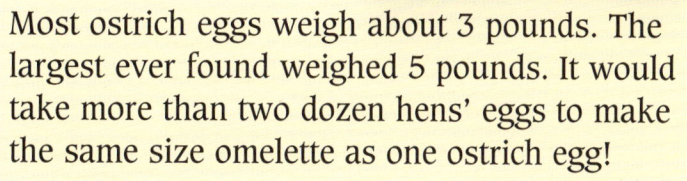

Most ostrich eggs weigh about 3 pounds. The largest ever found weighed 5 pounds. It would take more than two dozen hens' eggs to make the same size omelette as one ostrich egg!

The bee hummingbird always lays two eggs. Each egg is about the size of a fingernail and weighs much less than 0.1 ounce.

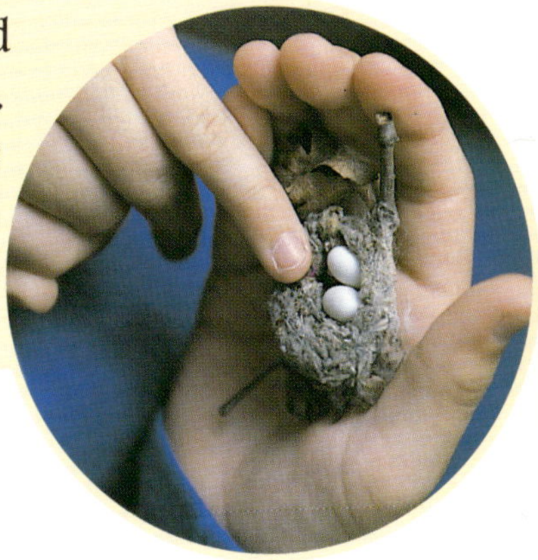

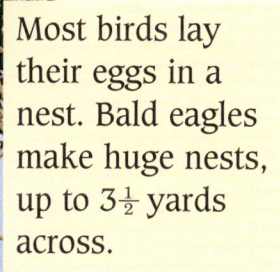

Most birds lay their eggs in a nest. Bald eagles make huge nests, up to $3\frac{1}{2}$ yards across.

The nests of some hummingbirds are less than $1\frac{1}{2}$ inches across.

1. How many of the hummingbird nests would fit across the bald eagles' nest? How did you figure out your answer?

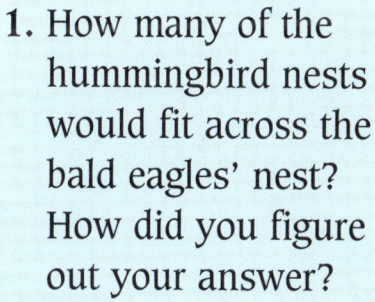

2. Choose one of the ounce weights on this page. How many of these would equal 1 pound?

Glossary

Longest, Tallest, Heaviest

animals
Living beings that can move about by themselves, have sense organs, and do not make their own food, but eat plants and/or other animals.

armspan
A length measured by a person's outstretched arms; the distance from the tip of one middle finger to the tip of the other. A person's armspan is approximately equal to his or her height.

balance
To be equal in weight, or to make two weights equal.

bird
A warm-blooded vertebrate that has two feet, two wings, and is covered with feathers. Birds lay eggs, and most birds can fly.

cubit
A measure of length used by several early civilizations. It was based on the length of the forearm from the tip of the middle finger to the elbow.

difference
In mathematics, the amount that is left after one number is subtracted from another.

dinosaurs
A group of reptiles that lived millions of years ago. The word "dinosaur" comes from the Greek words "deinos sauros" meaning "terrible lizard." It was first used in the form "dinosauria" in 1842 by Sir Richard Owen, an English scientist.

estimate
To make a general but careful guess about size, quality, value, or cost.

extinct
No longer living, having died out. For example, dinosaurs are extinct.

fossils
The remains, prints, or traces of plants and animals that lived thousands or millions of years ago. Fossils are usually found embedded in rocks.

handspan
The distance from the tip of the thumb to the tip of the little finger, when the hand is spread wide.

height
How high something is; the distance from the bottom to the top.

insects
Small animals. Adult insects have six legs and a skeleton on the outside. They usually have three body sections and two pairs of wings.

length
How long something is; the distance from one end to the other end.

mammals
Warm-blooded vertebrates whose young feed on mother's milk. Most types of mammals give birth to live young, but two, the platypus and the echidna, lay eggs.

primates
Members of the group of mammals that includes human beings, apes, monkeys, and lemurs.

reptiles
Cold-blooded vertebrates with dry scaly skin. Reptiles crawl on their bellies or creep on short legs. Snakes, lizards, alligators, and turtles are reptiles. So were the dinosaurs.

skeleton
The framework of bones of an animal's body. All vertebrates have a skeleton.

vertebrate
An animal with a backbone. Mammals, reptiles, and birds are all vertebrates. The dinosaurs were vertebrates; insects are not.

weight
The measure of how heavy something is.

wingspan
The distance between the tips of the wings when the wings are spread out. (Of birds, pterosaurs, planes, insects, bats, etc.)